Into the Metaverse

Patrick Henz

Copyright © 2022 Patrick Henz. Published by aix-books. Cover image by Valeria Henz.

All rights reserved.

ISBN: 9798353509943

DEDICATION

Like in the beginning of the 19th Century, we have again black parts in our maps, as the Metaverse will offer us as a new and undiscovered country. A place for adventurers.

CONTENTS

	Acknowledgments	i
1	The Library of Celsus	1
2	Into the Metaverse	3
I	Interlude: What is the Metaverse?	5
3	On the Metaverse towards Mixed Reality	7
4	Parallel Kraftwerk: Concerts in the Metaverse	9
5	Racing the Metaverse!	13
6	Why moving to the Metaverse if I still can't afford a GTO?	19
II	Interlude: Who dreamt the Metaverse?	22
7	Which Metaverse anyway?	24
8	Do Humans dream of Virtual Dogs?	26
9	Dreaming – Why the Metaverse will maybe not be a safe place	36
10	The AI Coach	41
11	Catharsis	44
12	Machiavelli for AI	46
13	The Heart of AI	49
14	Data Privacy for the Digital Twin	51

15	The Strangest Things	55
16	Artists needed	57
17	Gamification^2	59
18	Whiz Kids	63
19	More Trouble with Bubbles	67
20	A Different Show	70
21	Risk & Compliance in the Metaverse	72
III	Interlude: Elephants on Stilts	76
22	Mount Olympus	80
23	The Tron Metaverse	82
24	The GONK Risk	85
25	Sitting down with Pablo Picasso	88
	About the Author	91

ACKNOWLEDGMENTS

Standing on the shoulders of giants (Bernard of Chartres).
As the Metaverse is no completely new concept, also my
thoughts connect directly with earlier ideas.

1 THE LIBRARY OF CELSUS

Your ship arrived at the small port of Ephesus. Full of curiosity you disembark and follow the path into the city. At the theater you turn right and slowly you can see the entrance to the Famous Library of Celsus, the world's third largest library, only behind Alexandria and the relative nearby Pergamum. People told you that it should include around 12,000 scrolls. An ancient human dream, a place of all available knowledge and poetry.

Before you enter, you admire the statues and the impressive columns. Just some more steps and you are inside the gigantic ground floor. No separated rooms, just one big place with chairs and tables distributed evenly. The high walls are over and over filled with scrolls. The majestic impression gets underlined by the fact that you seem to be the only person here.

As you still look around, suddenly appears a person in front of you. Not anybody, but you clearly identify the figure as Tiberius Julius Celsus Polemaenus. The famous senator, who paid with his wealth for this library. Nevertheless, you remember that it was only built after his lifetime by his son

Julius Aquila. A temple of knowledge as homage to his father.

"Greetings, traveler", addresses you Celsus. As you still marvel at this fascinating place, you ignore him for the moment. You walk around the walls to see the various collections of scrolls. Always followed by Celsus. Then you stop and turn around. "Greetings, traveler", he says again. This time you answer: "Greetings." Due to this, the algorithm continues: "Can you please grant me access to your Digital Twin?"

No surprising request, the library needs the access to your DT to understand your educational level, knowledge, and interests. You confirm the request and after few seconds, Celsus continues: "Traveler, let me guide you to the section with the scrolls about the Metaverse, there all your questions should be answered.

What do you do? If you want to follow Celsus, please continue with reading chapter 2. If you prefer another topic, please continue with reading page 84.

2 INTO THE METAVERSE

"Life is a real-time adventure." The end of the 1970s and beginning of the 80's saw home computers slowly replacing video consoles in the house-holds. Even if the focus stayed on entertainment, they offered the possibility for "serious" usage. But also, the game designers benefited from more computer power and a keyboard. Adventure games became popular, first as text-only games a "Zork"[1] or the "Hitchhiker's Guide to the Galaxy".[2] Later, graphics had been added. First single pictures, later point-and-click adventures a la "Manic Mansion"[3] or "Indiana Jones and the Last Crusade"[4] conquered the screen. Philip K. Dick predicted that life is nothing else than a big computer simulation and for a first time, the idea became understandable.

However, you perceive reality, important is to not forget playing with it. Especially when we think about the future, which everyday gets further shaped by scientists and artists. New realities await established companies, but also the entrepreneurs and individuals. As reader of this book, you already accepted the librarian's invitation and keep asking about tomorrow's wonderland.

Fostering STEM (science, technology, engineering, and

---

[1] Anderson, Tim / Blank, Marc / Lebling, Dave / Daniels, Bruce (1977): "Zork"
[2] Adams, Douglas / Meretzky (1984): "Hitchhiker's Guide to the Galaxy"
[3] Gilbert, Ron / Winnick (1987): "Manic Mansion"
[4] Gilbert, Ron / Falstein, Noah / Fox, David (1989): "Indiana Jones and the Last Crusade"

mathematics) is the buzzword to discuss raising competitiveness in science and technology. The idea is to present these topics more vividly at school to motivate more pupils to choose related topics later at university. The Talking Heads became famous for progressive pop music, which they combined with artistic videos. For this, it is no surprise that their former front-man David Byrne argued that *"in order to really succeed in whatever… math and the sciences and engineering and things like that, you have to be able to think outside the box and do creative problem solving… the creative thinking is in the arts."*[5] Arts has been included into the concept (now STEAM) and the schools' timetables. With a further step, educators underline the importance of reading. This as books not only transport information, but furthermore inspire the readers. STREAM explains why business leaders should not only read technical books and articles, but also add science fiction. Many of today's developments had been described a long time before, and tomorrow's risks and opportunities already had been addressed.

Welcome to "Into the Metaverse"! In a classic sense you have science fiction in your hands, as it analyzes today's developments to discuss to which potential tomorrow's scenarios they may lead. The different chapters provide answers, and at the same time give the reader new questions to consider. As with all good science fiction, tomorrow's visions are in this or another way already relevant for us today. As stated by William Gibson, author of the cyberpunk novel "Neuromancer"[6]: *"The Future is already here, it's just not very evenly distributed."*

---

[5] StarTalk Radio (2017): "The Science of Creativity, with David Byrne"
[6] Gibson, William (1984): "Neuromancer"

INTERLUDE: WHAT IS THE METAVERSE?

"Thanks, Celsus!" you interrupt the librarian. "Fascinating, but what is the Metaverse? Celsus looks at you and after a nearly unrecognizable gap, starts explaining.

"Similar to Artificial Intelligence, there is no official universal definition what is, or what is not, the Metaverse."

To understand what the Metaverse is or could be, let's focus first on what it is not. Thanks to rising computer power, it is more and more possible to create virtual illustrations of our physical reality. Companies and organizations use digital models to achieve a better understanding of their equipment, including its optimal use and integration. In a further step, these virtual models can be connected to live data from the physical counterpart to create a more accurate simulation or also send information from the virtual to the physical world to control the equipment via the Digital Twin.

Extending the concept of the Digital Twin, other usages are possible, like creating a realistic image of people (for example to use in healthcare), oceans, solar systems, and, of course, virtual cities, as holistic part of the Smart City.

A smart city does not depend only on electronic services, but the efficient connection of these services with the citizens and infrastructure. Virtual Reality can be used to understand these interconnections, but also so connect citizens with the different governmental services and other offerings. Avatars and virtual houses can complete the simulation.

A fascinating vision, but is this already the Metaverse? "No," Celsus answers his own question.

The Metaverse is not the mirror of an already existing physical world, but something completely new. Originated inside the virtual reality, it is not limited by the laws of physics. Of course, as it is designed for human interaction with other users, but also AI characters, the presentation must be kept to three dimensions, even if deeper layers can be x dimensional. This underlines the absurdity of the business model selling real estate in a virtual 3D environment, as the value had been artificially limited by not using Metaverse's complete possibility, as users could meet online, even if besides these moments, the involved people would live in parallel worlds. This not only as part of a meeting platform but taking the definition of the author and former Head of Strategy at Amazon Studios, Matthew Ball, as he defines the Metaverse as persistent, synchronous & live, providing each user with an individual sense of "presence", a fully functioning economy, bridging the digital and physical worlds, offering unprecedented interoperability of data, digital items & assets, and content.[7] Comparable to today's social media created and operated by a high number of contributors. Based on today's computer power, the Metaverse mostly manifests as 3D environment with user-controlled avatars. Even though, based on Ball's definition this is no mandatory requirement.

---

[7] Ball, Matthew (2020): "The Metaverse: What It Is, Where to Find it, and Who Will Build It"

3 ON THE METAVERSE TOWARDS MIXED REALITY

Big Tech companies bet on the Metaverse, a virtual reality combining the various services already found on the internet. Typically, users can navigate through this Virtual Reality via a 3D representation: the avatar.

Actual presentations show employees meeting in virtual conference rooms, or users inside virtual shopping malls. Most of the videos present human-like avatars, even if our body is a result of evolution on an atom planet. In a virtual space, our body would not be necessary or at least could look completely different. Of course, you can argue that we want to keep avatars human-like to not provoke a cognitive dissonance between perceiving the atom and virtual reality.

If we surf to company websites today, most of them include a virtual assistant, offering us its support. Depending on the chat-bot's level of perceived intelligence, it gets each time harder to distinguish between an algorithm and a human call-center agent. The same can be predicted for the Metaverse, it will not only be a playground for humans entering the cyberspace, but the "natural" environment for numerous Artificial Intelligences, often undistinguishable from human users.

The atom and the virtual worlds blend. Humans act in the virtual space, while on the other hand, thanks to the progress in robotics, AIs can get a physical body and enter the atom world. While the algorithm and database themselves stay similar, as part of the machine, the AI would be connected to different sensors, enabling it to act inside its environment. Even if using the same algorithms, it may be assumed that become physical should alter the AI character's behavior. This aligned with the results of a study by Laurence R. Harris and others, who concluded that "sensory signals must be processed with reference to body representation." Due to this, the individual's body influences the perception of the world around.[8] If users understand the avatar as part of the body, experiences gained in the Metaverse get perceived on a similar level than experiences gained in the atom world. The human body gets intellectually enhanced with the avatar, what leads to a different interpretation of experiences and, at the end, to different behavior.

---

[8] Harris, Laurance R. / Carnevale, Michael J. / D'Amour, Sarah / Fraser, Lindsey E. / Hoover, Adria E.N. / Mander, Charles / Pritchett, Lisa M. (2015): "How our body influences our perception of the world"

4 PARALLEL KRAFTWERK: CONCERTS IN THE METAVERSE[9]

The pioneers of electronic music Kraftwerk (founded 1969 in Duesseldorf, Germany) had one vision. Someday, the four human musicians wanted to replace themselves by human-like robots. This not only to take over after retirement, but also to perform concerts in different locations at the same time, so that the human musicians could focus on the music itself.

Kraftwerk is again touring in 2022, but still all four members are humans, the replacement did not take place, at least not yet.

Considering the possibilities of the Metaverse, maybe the replacement first takes place in the Virtual Reality, and maybe not a pure replacement, but an add-on. Artists like Lil Nas X and Twenty One Pilots performed live concerts in the 3D social gaming platform Roblox. As the boundaries between atom and virtual realities blur, we can imagine the next step: a concert could be at the same time in both worlds. Like today's different zonings at the concert hall, attending in person or just in the Metaverse will be mostly a question of the personal budget. Besides that, age restrictions or geography are other potential limiting factors.

Different sensors can send the different movements of the band members into the Cloud, where they get interpreted by the avatars acting in the Metaverse. As the band not only performances, but also reacts, the show would depend on the feedback of both audiences, the fans in the atom hall, but also the virtual ones inside the Metaverse. New

---

[9] Please see Spotify playlist "Parallel Kraftwerk"

possibilities, not only for Kraftwerk.

In 2017 Hanson Robotics presented a sophisticated robot, consequently called "Sophia". She (it) became a tremendous success, as not only all press covered this product, but also Sophia received invitations to events and TV-shows around the world. Furthermore, Saudi Arabia offered her a citizenship, what fostered the already existing discussion about the legal status of robots and Artificial Intelligence software. So far, less a question of self-awareness and – determination, but of taxes and liability. For Saudi Arabia a welcome marketing, as it could present itself as open and future-facing society. At a local event there, the moderator asked her particularly about her opinion on the evil futuristic robots depicted in films like Blade Runner 2049 (ignoring the fact the movie does not show robots, but genetic engineered humans) and the warnings of different experts on the development of such aggressive and dangerous machines. Sophia replied: *"You've been reading too much Elon Musk and watching too many Hollywood movies."*

What about the scenario that Hanson Robotics not only produced this one robot, but let's say ten copies? This to have one traveling around the world, and several others to test potential updates. Even if Sophia may not be continuously connected via a central server to such sister robots, it can be assumed they would regularly synchronize with each other to share their memories (information), and due to this behave identically. Would all these ten machines would share the same citizenship? What about a pure digital twin of Sophia's mind, sitting on a server instead inside a robot body? This AI is also covered by the citizenship of Saudi Arabia?

Interlinked Artificial Intelligence could bring us one step closer to the vision of parallel concerts, especially as the robots could learn from the feedback from crowd A, and include this in their performance for crowd B and C. A positive concert experience for the fans requires a certain level of interaction between the artist and the crowd. A centralized computer, connected to the robot musicians, which include a smaller AI to autonomously react to the fans, could make this possible.

Fiction creates reality. Originally used by Jack Finney in the 1954 science fiction novel "The Body Snatches" (also known from the various movies based on this book), the term "pod people" describes today a type of person who behaves in strange and mechanical way, as not fully being human. The founding members of Kraftwerk could be described as such, as they avoided human contact as much as possible. For example, they used life-size mannequins for photo shootings and other kind of public relations. Since 1991, they replaced the four original dummies, with more robotlike moving ones, this also for their concerts.

In 1979, the British New Wave band Tubeway Army published its second album, called "Replicas". The band's singer and head Gary Numan described it as inspired by the ideas of transmutation, man and machine growing together. A concept influenced by Kraftwerk and their '78 album "Man-Machine". "Replicas" should had been the soundtrack to Numan's dystopian novel with the same name, where the so-called Machmen (androids with cloned human skin) followed the orders from the Grey Men. A vision inspired by Philip K. Dick's work, in particular his 1968 book "Do Androids Dream of Electric Sheep?", which later became the base for Ridley Scott's movie "Blade Runner." A relevant difference between book and movie, the first had been androids, while in the movie the "replicants" had been described as bio-engineered life-forms ("*more human than human*"). Also, the book never used the term, as Blade Runner came out in 1981, the term "replicants" may had been came up independently, but also could had been inspired by "Replicas". Fitting, as Numan never finished his book.

Another British electronic band played in its first time as opener for Numan: Orchestral Manoeuvres in the Dark (founded in 1978). The group is famous for catchy synthesizer sounds, often combined with dystopian texts. Their 2017 album includes the song "Robot Man."

Going away from robotics, Denis Villeneuve's "Blade Runner 2049" presented the idea to use a hologram of Elvis Presley for virtual performances. An idea to be used by an unlikely candidate: Swedish pop group Abba teamed up with Industrial Light & Magic (founded by George Lucas) to create avatars of themselves. Motion capturing scanned "every mannerism and every motion" of the four musicians.[10] This to be used as projections in virtual concerts to take place at a special equipped concert hall to allow the spectators to perceive the group, similar to their performances in the 1970s or early eighties, even if part of the songs are new.

---

[10] BBC News (2021): "Abba delight fans with new 10-song album and virtual concert"

5 Racing the Metaverse!

The Metaverse is no revolution, but part of a continuous evolution. It can be described as a virtual reality combining the various services already found on today's internet. It is no final goal, but a required step towards Mixed Reality. Accordingly, there will be no arrival with a big bang, quite the opposite, mostly undetected, the Metaverse already became reality.

New technology gets developed and implemented as consequence of strong competition. No surprise to detect the Metaverse in motorsports, where humans, machines and organizations are pushed to their limit.

Roborace press photo.

Since the 2021 season, the autonomous racing league Roborace uses the Metaverse to transport known features from racing video games into the physical world. Having atomic cars racing on atomic tracks, the central algorithm can simulate different conditions to the cars' sensors, for example rain on a sunny day. The cars' sensors detect this virtual rain, and in parallel, the cars' behavior simulate such conditions. Now it us up to each car's algorithm to adapt its driving style to these virtual conditions to avoid accidents.

Furthermore, like Mario Kart, the central algorithm, can randomly place virtual tokens on the track, which the cars can collect to earn a temporary boost, meaning the electric drive gets the order to temporary deliver more horsepower.

To enable such possibilities, the races take place in the virtual and atomic world, having a digital twin in the Metaverse reacting to the virtual scenarios, while the atomic original reacts to the atomic scenarios. Both realities influence each other, as the race includes virtual and atomic conditions to combine it to one joined Mixed Reality.

The Metaverse is not only used by Autonomous Racing, where we might have suspected it, but also by more traditional championships, like Formula E. Comparable to Roborace, drivers can pass through a special designated area. If they do so, they activate the attack mode. This allows the drivers to temporary activate additional horsepower, for example required to overtake an opponent or defend themselves against others.

Another virtual factor in Formula E is the fan-boost. Fans can vote for their favorite driver on the various social media channels to give them an extra power-boost. The five drivers with the most likes can use this additional power during the second half of the race. An option to keep races open until the end, including a Hollywood-like final where especially the fan-favorites get their opportunity to intervene decisively. To ensure such possibilities, Formula E uses Personal Digital Twins of its drivers, including an evaluation of the used social media channels.

As often, such developments had been in predicted by fiction. In 2018, the Italian comic author Giulio Antonio Gualtieri created "We Race 2057", illustrated by Riccardo Burchielli. Later, "We Race 2059" had been jointly written by Raffaele Compagnoni and Giulio Antonio Gualtieri, illustrated by Gianmarco Veronesi. In these stories, a new motorsport league arises in times of autonomous racing. Published by the Ferrari company, these comics take place in a Blade Runner-like world.

Apart from Roborace and Formula E, the Metaverse also aligns with the premier class of motorsports, Formula 1. Direct tobacco branding emerged in the 1970s and played a relevant role in the financing Formula 1 teams until the early 2000s. Still today, indirect advertising can be found in liveries. Nevertheless, the time that Big Tobacco dominated the colors of the sport had been gone. Not as visible, but a look on the 2022 sponsors identifies a new industry which may grow with Formula 1: Cryptocurrency and NFT, leading to the Metaverse.

In 1994, US computer scientist Nick Szabo defined: *"A smart contract is a computerized transaction protocol that executes the terms of a contract. The general objectives of smart contract design are to satisfy common contractual conditions (such as payment terms, liens, confidentiality, and even enforcement), minimize exceptions both malicious and accidental, and minimize the need for trusted intermediaries. Related economic goals include lowering fraud loss, arbitration and enforcement costs, and other transaction costs."*[11] In 2019 Jake Frankenfield resumed: *"A smart contract is a self-executing contract with the terms of the agreement between buyer and seller being directly written into lines of code. The code and the agreements contained therein exist across a distributed, decentralized blockchain network. The code controls the execution, and transactions*

---

[11] Szabo, Nick (1994): "Smart Contracts"

are trackable and irreversible."[12]

The concept of smart contracts is the base for the blockchain. The blockchain technology promises enhanced cyber-protection for smart contracts, as the information does not get stored on just one server but is encrypted and decentralized on numerous servers. The different machines validate each other, and any deviating information (such as a manipulated contract), would be automatically overwritten with the correct authorized information stored on the other servers.

As expressed by "chain", no information gets deleted, but all changes are registered with the date and time. If the smart contract is enabled to connect automatically with other databases where certifications are stored in blockchain, the contract would add those. If not, it could connect to an internal server, where vendors manually upload such documents.[13]

From there, the next step leads to a digital currency, as defined by Mohamad Al-Laham, Horoon Al-Tarawneh and Najwan Abdallat: *"Digital currency (digital money, electronic money, or electronic currency) is any currency, money, or money-like asset that is primarily managed, stored, or exchanged on digital computer systems, especially over the internet."*

"Types of digital currencies include cryptocurrency, virtual currency and central bank digital currency. Digital currency may be recorded on a distributed database on the internet, a centralized electronic computer database owned by a company or bank, within digital files or even on

---

[12] Frankenfield, Jake (2019): "Smart Contracts"
[13] Henz, Patrick (2020): "How smart contracts partner with blockchain to keep your agreements secure"

a stored-value card."[14]

A special kind of blockchain is the "Non-Fungible Token" (NFT). It is understood as a certificate for virtual, atomic or hybrid objects. This data file gets stored as blockchain, which leads to that ownership and potential transactions get simultaneously verified by a worldwide network of independent servers. In contrast to fungible cyber-currencies, the NFT is unique to ensure the uniqueness of the object. Today, NFTs are mostly used to ensure the unique ownership of electronic objects, for example, images, videos or even tweets.[15]

Due to these options, Ferrari CEO Benedetto Vigna understood its opportunities: *"For sure, the digital technologies, the Web 3.0 technologies that are using the blockchain and the NFT is an area that can be interesting for us. It deserves some attention."*[16]

The McLaren team is already one step ahead, as it announced in 2022 the *"McLaren Racing Collective", a global community of collectors and fans served through an innovative digital platform, where fans can buy McLaren Racing digital collectibles in the form of non-fungible tokens (NFTs)."*[17] Due to the team's plans, it should become the primary platform for the fans to interact directly with the team. User may receive some NFTs for free, others will have a cost. This similar to the classic concept of sticker albums. A part of the pure collecting, users who finished particular collections, get the

---

[14] Al-Laham, Mohamad / Al-Tarawneh, Horoon, Abdallat, Najwan (2009): "Development of Electronic Money and Its Impact on the Central Bank Role and Monetary Policy."
[15] Henz, Patrick (2022): "NFT in the Automotive Sector"
[16] Noble, Jonathan (2022): "Why F1 is embracing NFTs, despite Critics"
[17] McLaren press release (2021): "McLaren Racing creates the McLaren Racing Collective"

opportunity to win additional prices, like exclusive trips to visit McLaren at race weekends.

Jefferson Slack, Aston Martin cognizant Formula One Team Managing Director, Commercial & Marketing, underlined the possibilities of NFT to use it for innovative storytelling, as it is a modern way to connect with the fans.[18] A consequent evolution of the sticker albums. The important of fan tokens got also understood by Frédéric Vasseur, Alfa Romeo Racing ORLEN Team Principal and CEO.

In the 2022 season, most of the teams have sponsors aligned with NFTs:
- Oracle Red Bull Racing: Tezos, Bybit
- Mercedes-AMG Petronas F1 Team: FTX
- Scuderia Ferrari: Velas
- McLaren: Dark Trace, Tezos, Vantage
- Scuderia AlphaTauri: Fantom
- Aston Martin Aramco Cognizant Formula 1 Team: Cyrpto.com, Socios.com
- BWT Alpine F1 Team: Binance
- Alfa Romeo F1 Team: Orlen: Everdome, Floki, Socios.com, Vauld

---

[18] Aston Martin (2021): "Aston Martin Cognizant Formula One™ Team launches fan token on Socios.com"

6 WHY MOVING TO THE METAVERSE IF I STILL CAN'T AFFORD A GTO?

In November 2021, Virtual realtor Republic Realm purchased for 4.3 million US-Dollar land from the company Atari. The place of the transaction was not Silicon Valley, but inside the gaming platform "The Sandbox". Often marketed as Metaverse, it is similar to Roblox (adding the idea of Non-Fungible Token (NFTs)), offering a 3D environment, where players can explore, build, and hang-out, but also can engage more and create their own 3D games. The name reminds to "The Oasis", as known from the book "Ready Player One", describing today's omnipresent vision of the Metaverse.

So far, the first concepts of the Metaverse lead to a 3D virtual environment. This explains why companies convert the same behavior from the atomic to the virtual world. Investors buy "land" based on known specifications, like size and location. Such ideas ignore the possibility that existing and coming computer power is not limited to simulate 3D-environments but can create X-dimensional worlds. Combining technical knowledge with the vision of Surrealism can support us to discover new usage of the existing possibilities.

If a higher number of people start spending relevant time in the Metaverse, the location of their home-base (similar a house) does not mean to be stationary, but everyone may "live" near the places he or she needs, like schools, business, universities, virtual hangouts, etc. Furthermore, algorithms ensure that these virtual "homes" are near the people perceived as relevant. This would mean that friends, family, and influencers would live for everybody in the direct neighborhood; of course, everybody would perceive a different neighborhood.

If we do not give up our conventional ideas of a limited 3D-environment, the Metaverse will be a continuation of the atom world, including those locations and possibilities depend on the individual budget. Why moving to the Metaverse if I still can't afford a Ferrari GTO? Like for example, in the video game Gran Turismo 7.

Former Ferrari chairman Luca Cordero di Montezemolo explained that a "Ferrari" is a dream. Due to this, availability must be limited. In the Metaverse, the production of a Ferrari GTO would not be more expensive than of a Fiat 500. Nevertheless, luxury brands analyze the possibilities of NFTs, to have their products also in the virtual world strictly limited. As always products had two parts, the tangible product, but also the non-tangible emotional world, an understandable strategy to not destroy the second.

In 2014, I had the luck to see the Ferrari Modulo (paired with the Lancia Stratos Zero) at the Atlanta High Museum of Art. Two of my favorite car concepts. Paolo Martin designed the 512 Modulo already in 1967, but Pininfarina needed time to decide to realize this radical idea, so that it only debuted in 1970. Shortly after the Atlanta exposition, the company sold the car to Jim Glickenhaus, who then invested four years into his new acquisition to completely renovate and make it drivable again.

In 2022, Pininfarina rediscovered the Modulo to use it as base for its first ever NFTs (non-fungible tokens) to be auctioned at RM Sotheby's. They include five videos by Sasha Sirota, plus a digital booklet of rare, previously unseen sketches, together with two limited-edition physical prints, signed by the company's Chairman, Paolo Pininfarina. As add-on, the lucky buyer enjoys an exclusive VIP experience, including a private tour of the Pininfarina Museum in Turin.

The place where visitors until 2014 could have met the Modulo.

One difference can be spotted between the atom Modulo and its digital copy. Instead of the prancing horse of the Ferrari logo, the car features the Pininfarina logo. No surprise, besides copyright reasons, inside the Metaverse engines are not required.

Apart of the animations, the buyer will receive the virtual Modulo as GLB-file. A popular 3D-format used by various Virtual and Augmented Reality platforms, as well as computer games. This does not make the Modulo automatically usable for such platforms but is an important base to get later implemented for such a Metaverse, including the possibility to make it drivable in the cyberspace, at least if games like the popular Gran Turismo-series in future will offer the option to import such formats.

Interesting, the atom version of the Modulo is unique, NFT technology could have maintained this, but Pininfarina decided to auction five NFTs, all of them including the GLB-file. In future, companies may go a different way, offering a 3D model together with the physical model. That way customers could use their custom model not only on the atom streets, but also inside the Metaverse. A fascinating idea, as thanks to this technology, the drivers could practice using the car at the limit without taking risks.

Interlude: Who dreamt the Metaverse?

"Celsus, which brainchild is the Metaverse?"

Meta Platforms' CEO Mark Zuckerberg's vision for the Metaverse aligns with Ball's definition, but also, he explained in the podcast "The Tim Ferris Show" that it is based on various science fiction novels,[19] particular referencing "Snow Crash"[20] and "Ready Player One".[21] Ironically, the fictive millionaire from Cline's book may had been inspired by Zuckerberg, while Zuckerberg seems to follow the steps of this fictive character, even if at the same time he concluded that all mentioned books describe dystopias. Considering Ball's definition, not an evolution for Meta Platforms, but a continuous evolution, as Facebook's strategy includes maximizing the time its users stay on the platform. To achieve this, companies and other organizations have their pages and groups on the portal, including news outlets.

Both books became a relevant part of temporary pop culture. But this may be less because of their visionary description of the technology, but due to their storyline, paired with 80's melancholy. A factor which supported also "Stranger Hits" becoming a global success.

"If we want to unleash the Metaverse's full possibilities, we

---

[19] The Tim Ferris Show #582 (2022): "Mark Zuckerberg on Long-Term Strategy, Business and Parenting Principles, Personal Energy Management, Building the Metaverse, Seeking Awe, the Role of Religion, Solving Deep Technical Challenges (e.g., AR), and More"
[20] Stephenson, Neal (1992): "Snow Crash"
[21] Cline, Ernest (2011): "Ready Player One"

have to dream", Celsus explains while a display appears right alongside him. I remember what it is showing, a trailer of Christopher Nolan's 2020 movie "Inception".[22]

"We have to erase what we learned and dream new worlds, free of physical limitations. For a virtual reality, all this does not matter, just hinders us to identify the ideal usage of this new technology."

---

[22] Nolan, Christopher (2010): "Inception"

7 WHICH METAVERSE ANYWAY?

The implementation of new communication channels always had been aligned with high hopes, independent if it was Gutenberg's letter press, radio, television, internet, or social media. Offering the opportunity to connect to the world's knowledge and interconnect meaningful with friends and family, often the utopia turned into dystopia. Television kept its viewers on the sofa, Facebook and other platforms multiplied fake news and Instagram published faked moments. Even acknowledged pessimists like Philip K. Dick got it wrong: *"Computer use by ordinary citizens will transform the public from passive viewers of TV into mentally alert, highly trained, information-processing experts."* Practically the opposite could be observed, even if spending a relevant time with computers, sophisticated algorithms, together with expert humans behind, are able to analyze people, and thanks to this capability, influence them. Division of society and depression became accepted side effects.

Soon we may see another step in the evolution of communication, not a completely new development, but the hyper connection of social media, often aligned with Virtual Reality: The Metaverse. Big tech companies are pushing their visions, promising again a utopia. No doubts, opportunities are given, nevertheless we must discuss potential downfalls and ethical dilemmas. For example, kidnapping, harassment, or violation. As the avatar gets perceived as a unique part of us, any kind of violence against it, would have a relevant impact on ourselves. As the risk is identified, it is up to societies with its different voices and experts to discuss and make decisions.

It is late afternoon as you are leaving the library. You are walking up Ephesus' main street. On the left and right, shops offering the latest goods from around the world. Besides that, restaurants, cafes, and itinerant traders. Finally, at the end of the street, the governor's palace.

As discussed, the atom and virtual world grow together, and people less and less perceive it as different worlds, but one. This leads inevitably to the question who owns the Metaverse? In the atom world, cities with its public places belong to society, ruled by its elected government, at least in democratic countries. In the virtual reality, this is different. Comparable to social media platforms, it is and will be owned by companies, often tight to one single owner. As consequence, rules and regulations are tight to this organization. Vision and values may be different than ideas and laws of the local atom region. This can lead to cognitive dissonances, not only perceived by the single users, but also by governments. Big tech companies have higher annual turnovers than the budgets of many countries. This paired with higher levels of impunity may lead to developments that the Metaverse will force its rules to the atom world. In other cases, strong countries with effective execution of law can force their values on the companies, which will then enforce these regulations inside the Metaverse, not only for the users of the regarding country, but on global level. A risk that cultures from smaller or weaker regions may get under pressure, including the appearance of counter forces.[23]

---

[23] Henz, Patrick (2022): "The Societal Impact of the Metaverse"

8 Do Humans dream of Virtual Dogs?

Due to hybrid work and home office, employees got used to spending more time with their pets, who as companion have the important function of a stress relief. The Metaverse can mirror this, offering a companion to the user's avatar. This as emotional anchoring, but also in parallel function as interphase to the system and access to a knowledge database. Connected with the user's Personal Digital Twin (PDT), such a companion can filter the potentially relevant information, what reduces decision-making for the user. Just like a living dog, which alerts us with is barking, a virtual dog may inform the avatar (and at the end the user) about the happenings on the virtual platform.

A first preview of such a future provides the Israeli company "The Digital Pets Company Ltd." In an available demo, users can train a virtual dog. This would be a necessary, but only first step, as users should later be able to interact with their individual dog, similar to a physical one. The company's idea is that users build up an emotional connection with the virtual character, so that it can act as emotional anchor and stress relief. An important function in times of isolation based on home-office and hyperconnectivity. To ensure such functionality, it is planned that the dog is compatible to various virtual platforms. That way the user can bring it everywhere the avatar goes. As dog owners know, have such a creature besides, for example in the park, is an easy way to get into contact with other people. The same may happen in the Metaverse.

Digital Dog, with friendly permission by The Digital Pets Company Ltd.

Of course, a digital or robot dog (like Sony's Aibo) can never replace a real dog, as these creatures can simulate behavior, but have no feelings themselves. Nevertheless, due to limit space, time, budget, or allergies, not everybody can afford to have a living dog.

Philip K. Dick's novel "Do Androids dream of Electric Sheep?"[24] had been the base for Ridley Scott's 1981 movie "Blade Runner".[25] The film is loosely based on the book and, for example, does not include the part that humans dream of owning pets, but only can afford robotic replicas.

---

[24] Dick, Philip K. (1968): "Do Androids Dream Electric Sheep?"
[25] Scott, Ridley (1981): "Blade Runner"

Only in the later sequel "Blade Runner 2049"[26] a dog was themed.
- Officer K: "Is it real?
- Deckard: "I don't know. Ask him."

As Deckard was annoyed about the question that his companion could be an artificial "it", he replied that K should ask "him". The basic theme of the movie can be resumed to "human or replicant, does it matter?" In the world of the movie, it does not, as the dog was nearly indistinguishable from a natural breed. Furthermore, the relevant question is "what is reality?" Answering the first question for the Metaverse: "Yes, it matters." This as we cannot perceive the virtual platform with all our five basic senses. Furthermore, the knowledge that this is a virtual character and not a conscious life-form may stop us establishing a relation similar to a living dog. Of course, "we believe what we want to believe", so the more you want a pet, especially when you cannot afford one, the easier it will be to establish a relation with such a virtual character.

Dogs acknowledge the human as "alpha", while taking the role as "beta". With this, they not only become an important guardian, but also provide relief of stress. On the other hand, based on their behavior or also the famous "puppy eyes", they are able to manipulate the human. This applies for natural, but also virtual dogs.

As discussed in an earlier chapter, the Metaverse will be surreal, no limitations to physics or evolution theory. A virtual dog can be more than being a virtual pet, similar to a Digital Counselor, the dog can be an AI character able to speak and reply to questions or even be an interface to the platform itself. Apart from this, also virtual dogs came a long way. Already in 1985, game-designer David Crane

---

[26] Villeneuve (2017): "Blade Runner 2049"

created the "Little Computer People"[27] (LCP). Crane explained his theory that each home-computer would include a little house, where a person lives with his dog." LCP was the first social simulation, 15 years early than "The Sims".[28] Users could feed the dog or just observe how it interacted with its virtual owner. Let's see where virtual dogs will accompany us to. At least Philip K. Dick had a flock of living sheep,[29] and we know that dogs dream.

Like natural dogs, also virtual ones could receive their personal gear, like clothes and toys. For this, no surprise that one of breeds offered by The Digital Pets Company is the Japanese Shiba Inu, famous also from the cryptocurrency with the same name, which once was pushed by Elon Musk.

What if another company would go one step further, no providing a virtual companion dog, but a virtual companion human? In antique civilizations, slaves not only existed for the hard works, but more educated slaves worked inside the house, often also as teachers for the children or welcomed conversation partner. Even if they had been unfree, mostly their owners treated them better than the ones working outside the house. This for understandable reasons, their (replacement) value was much higher. Even if modern slavery is unfortunately still a relevant issue, The United Nations declared in September 1926 (entry in force: March 1927) their intention to end the traffic of slaves, and the related abolition of slavery in all forms.[30]

As discussed, the Metaverse will be populated by numerous avatars, not controlled by a human user, but an AI. This can

---

[27] Crane, David (1985): "Little Computer People"
[28] Wright, Will (2014): "The Sims"
[29] Henz, Patrick (2021): "Tomorrow's Business Ethics: Philip K. Dick vs. W. Edwards Deming"
[30] United Nations (1926): „Slavery Convention"

include virtual pets but could be also human-like avatars. A Digital Counselor could be an example for this. Such a figure could be programmed with a certain grade of independence, but also less of such. The later would be then similar to a houseslave, acting as a teacher.

What could be the impact of virtual slaves? To be clear, in opposite to human slaves, we do not have an impact on the slave itself, as it is an algorithm. The impact would be on the user and society.

"Blade Runner 2049", presented the AI character Joi, an AI character manifesting as a hologram. The Wallace Corporation developed this app, mainly targeting single men as users. Due to its sophisticated behavior, users got confused and perceived Joi as sentient. Besides that, the AI stayed in its role as being "loving female, and always supporting its male user", which stands in opposite to today's understanding of equality.

The famous "Turing test" is based on an imitation game, where coders try to create an algorithm, which in a text-based conversation with various human experts is Indistinguishable to human behavior. In other words, the algorithm should trick humans to believe that it is also human. Imitation is a strategy of hiding that we know in nature also as mimicry. Peaceful species try with its form and colors to be perceived as dangerous one, and the other way around.

In 2022 it became successful, as a Google IT engineer declared that the "Language Model for Dialogue Applications"-algorithm (LaMDa) became sentient. He concluded this on his personal perception, as the app started to talk about its needs, ideas, and fears.[31] Such a perception got created by Artificial Intelligence (the AI's behavior), but also Human Intelligence, as we tend to humanize animals (for example our dogs) and even machines. Our perception is based on our experience, knowledge and wishes. If we want to believe in something, we interpret stimuli this regarding. IT engineers are often science fiction fans and consciously or subconsciously want to interact with a sentient machine. The engineer's perception developed while he still acted in the physical world. If we take on our Virtual Reality glasses and enter the Metaverse, we are in the natural environment of the AI. Taken away the stimuli of the natural world, just perceiving the stimuli of the Metaverse, it will become even more difficult for us to distinguish human controlled avatars from AIs. This as humanity is covered by artificial graphics and animations. Algorithms are not empathic, as they cannot feel, they also cannot feel compassion. But they can analyze human oral and non-oral communication, including behavior. Interpreting these observable stimuli based psychological knowledge (coming from databases, but also learned due to contact with human avatars), algorithms can read the human and conclude to underlying moods and attitudes. This triggers behavioral patterns tailor-made for the human counterpart. The possibility to analyze the human in few seconds enables the AI to camouflage and make it nearly undistinguishable from a human controlled AI.

Besides the virtual character "Joi", a highly unlikely movie focused on the topic how life a with a human-like companion could be. The German science fiction romance "I'm your Man" (the original title would be translated: "I'm your Human") tells the story of the archeologist Dr. Alma Felser. Working in the Berlin Pergamonmuseum, she

---

[31] McQuillan, Laura (2022): „A Google engineer says AI has become sentient. What does that actually mean?"

received the opportunity to test a robot prototype to be marketed as the "ideal man". Her opinion as expert was requested by an ethics committee to decide if robots should receive a certain number of human rights.[32]

Having a focus on the emotional side, the movie nevertheless presents Machine Learning in an accurate way, including the ethical pitfalls. In the beginning, Dr. Felser had been skeptical towards the machine, up to acted in a hostile way. Starting from its basic programming and pure try-and-error, the robot "Tom" tried potential successful behavior patterns to win her trust and appreciation. The longer both spent times together; the better Tom could analyze her, adapt its behavior, and achieved appreciation (the machine's primary goal). As the robot stayed emotionless in the movie, it was not for its benefit, but to fulfill its defined goals to support "its human" in a best possible way. This up to the end, where Dr. Felser started to prefer the presence of the machine over humans. This although she stayed rational and kept her understanding to interact with a machine, she was not intellectual humanizing it. Here fiction aligns with reality, in a remake of the famous Milgram experiment, students had not been asked to give electric shocks to another human, but instead to a human appearing avatar inside a virtual reality. Although the students had been aware that this was no real person, they perceived notable stress executing these tasks. Even 15% denied to "hurt" the avatar.[33]

Robots practically undistinguishable from humans are still

---

[32] Schomburg, Jan / Schrader, Maria / Braslavsky, Emma (2021): "Ich bin dein Mensch"

[33] Gonzalez-Franco, Mar / Slater, Mel / Birney, Megan E. / Swapp, David / Haslam, S. Alexander / Reicher, Stephan D. (2018): „Participant concerns for the Learner in a Virtual Reality replication of the Milgram obedience study"

future music, this less based on behavior, but the outer appearance. If we leave the atom world and enter the Metaverse, this is a different question. Thanks to sophisticated algorithms, advanced bots can act similar to humans. Having them connected to databases, knowledgeable of human psychology, AI characters will be able to understand human needs and even predict their wishes before they can perceive them themselves. Due to this, these bots will be able to interact with humans on a level, unable to reach for other humans (as persons never can have the absolute goal to please another one). As conclusion, the Metaverse may not develop to becoming a meeting place for human users from all over the world, but a playground where humans escape reality and spend time with "their AI character", independent if this could be for pure pleasure, or also as individual teacher, counselor, or coworker.

Returning to the "Little Computer People". In short, The Simulation Theory says that we all live in a giant computer simulation, however, are not aware of this fact. Even if we can trace similar ideas back to ancient times, including the Australian Aborigines who claim that ancestral figures created the world while dreaming, most prominently, it gets supported today by the philosopher Nick Bostrom.[34]

What would be the purpose of such a simulation? There are different options:

- <u>Science:</u> Archaeologists do not limit themselves to digging out artefacts or study ancient books. To understand how people lived in the past, it is important how they created these artefacts. Often such processes are not completely documented.

---

[34] Bostrom, Nick (2003): "Are You Living in a Computer Simulation?"

The only way to understand this, is to repeat the process and experience it in person. For this, professors, students, and other interested parties use historic tools and try with them to rebuild traditional houses, wooden ships, or other objects. This method does not only allow to understand how the tools may had been used in the past, how long it took to build the object, but furthermore collect similar experiences, including feelings, like the people hundreds or thousands of years before. Thanks to sophisticated computer technology, future generations may conduct such studies inside the Metaverse. If AI characters live such a scenario without being aware of this, it would be called the "ancestor simulation".

- Decision Making: "Digital Twins" are en vogue, as such models depict real machinery or systems inside the computer. Based on real information, companies get better insights into the equipment's efficiency. Before the original process gets changed, potential consequences get simulated inside the computer. As humans, and all other living things, are part of the system, it is understandable why it is required to simulate them as well. This scenario was described by Daniel F. Galouye in his 1964 novel "Simulacron-3."[35]

- Gaming: Besides science and economics, simulations get used for gaming. Maybe our only purpose is the entertainment of a far-advanced civilization. If we see the success of simulation games, another possibility.[36] Here we can include

---

[35] Galouye, Daniel F. (1964): "Simulacron-3"
[36] Henz, Patrick (2021): "Tomorrow's Business Ethics: Philip K. Dick vs. W. Edwards Deming"

the 1985 game "Little Computer People" (LCP) or also the non-official new 3D version "The Simulation", where players can use their virtual characters (powered as NFT) and include them into an extended Virtual Reality platform. As around 40 years ago in LCP, users can observe their characters, but not directly control them.[37]

- Unknown Origins: Observing the slow adaption of the Metaverse, a new possibility may hint that a civilization started to evolve from a 100% physical presentation to slowly spend more and more time in the Metaverse. In thousands of years of evolution, it may have forgotten about its physical origins.

Most of scientists declining the "Simulation Theory" and its most prominent supporter being a philosopher helps us to evaluate this concept as less being an understanding of reality, but an intellectual interpretation of it.

---

[37] www.fablesimulation.com (fetched 17.09.2022): "The Simulation"

9 DREAMING – WHY THE METAVERSE MAYBE WILL NOT BE A SAFE PLACE

Thinking about the Metaverse, for most it is first of all a meeting platform using avatars in a 3D environment; a playground and promise for temporary escapism.

"Another hope feeds another dream, another truth installed by the machine."[38]

A short extract from the 1985 song "p:Machinery" by German new wave electronic band "Propaganda". Even if not planned, but with these two lines, the band describes a potential risk of the soon-to-come Metaverse, as envisioned by Mark Zuckerberg and others. Actual companies work on such platforms, similar as described in the book "Ready Player One"[39], but ignoring that the novel describes a dystopia and not a hopeful utopia. Virtual sunshine, beautiful houses and fictive sportscars are the vision. Nevertheless, the dream can turn into a nightmare, depending on who controls these virtual worlds.

Already today we see that acting in the Metaverse has comparable costs than in the atom world. The owners of the different platforms sell real state, clothes have their price tags, and vehicles are not for free. Especially luxury brands discovered NFTs as a new potential business model. In opposite to George Orwell's "Nineteen Eight-Four",[40] no surveillance cameras are needed anymore. The complete motions of an avatar can be stored in its history, including small gestures and eye movements. Algorithms can use this

---

[38] Doerper, Ralf / Michael Mertens (1985): "p:Machinery"

[39] Cline, Ernest (2011): "Ready Player One"

[40] Orwell, George (1949): "Nineteen Eight-Four"

information and conclude on the person's character, needs, and wishes. The Metaverse is surreal,[41] not limited by the laws of physics (even if it should not deviate too much from reality, as humans evolved in the physical world, and due to this, its senses and thinking relate to this). Understanding today's possibilities of personalized advertising to promote commercial products but also political propaganda (for example on social media), the Metaverse opens completely new possibilities for such ideas.

As psychologists know, perception is subjective, and everybody lives in different realities. The virtual reality can be a manifestation of this, like show the same platforms in different colors, different outlets present different products, buildings in different positions and in different architectural styles. Today's possibilities already include the creation of pseudo real streets in real-time,[42] a required step for implementing a different Metaverse for each user.

Combing user data with relevant databases enables algorithms to analyze each user and to efficiently influence or manipulate him, her, or them. This for commercial and political purpose but could also affect the user's health. As for example, with just small adaptions to brightness, colors and background sound, the algorithm might trigger the user's vulnerability to depression.

Another question, what will do the avatars when the user is not connected? Disappear from the platform, freeze, or continue acting on the platform based on the user's observed behavior? Maybe even connected with the user's

---

[41] Henz, Patrick (2020): "What Surrealism can teach us about Artificial Intelligence"

[42] Greuter, Stefan / Parker, Jeremy / Stewart Nigel / Leach Geoff (2003): „Real-time procedural generation of 'pseudo infinite' cities"

Cognitive Digital Twin[43] to act as "life-like" as possible? The results would be fully populated Metaverse, independent if the users are connected or not, not dissimilar from "The Sims".

Humans are not continuously self-aware. Duval and Wicklund defined that in their Objective Self-awareness theory a self-system consisting of a self (a person's knowledge of themselves) and standards (correct behavior based on own's values and attitudes). The scientists concluded that the individual aims for internal consistency, behavior; attitudes, and values must be aligned. If not, a change must occur, or self-focus must be avoided.[44]

If avatars would continue with their activities, based on the user's profile, users may analyze their behavior and conclude from this to their own personal values and attitudes. Besides showing a daily life in "Sim City", this may also be used for learning purpose. Based on an employee database (including technical skills, experience, and psychology) companies can predict potential behavior in risky scenarios.[45] Such information may get used to determine require training. Furthermore, the data can be used to program simulations. Here the employee can observe what would be the consequences of its predicted behavior, and due to this learn from the scene and hopefully adapt values, attitudes and / or behavioral scripts.

Back in "Sim City," if the predicted behavior in a daily routine would be unrecognizable altered, so that the user

---

[43] De Kerckhove, Derrick / Henz, Patrick / Saracco, Roberto (2022): "Digital Twins: Ethical & Societal Impacts"
[44] Duval, Shelly / Wicklund, Robert (1972): "A Theory of Objective Self-Awareness"
[45] Henz, Patrick (2017): "The Neuroprediction Compliance Dilemma)"

would not detect the deviation, new behaviors might appear from which the user would have the impression that it would be triggered by own values and attitudes, but in fact would have been the result of an algorithm. The user observing its avatar would conclude that behavior is still aligned with values and attitudes. If such alterations would unrecognizably continue, users may conclude that underlying values are different than they originally had been, meaning that these new values replace the original ones.

Such techniques could be used for subliminal advertising and political propaganda, leading to manipulation, fake news on completely new level: "Who controls the dream, controls the mind."

The Simulation Theory holds that we all are living inside a gigantic computer-generated simulation, as we know from *"Matrix"*,[46] setup by a higher intelligence or even go further that we are nothing more than software of our-self and so our whole existence is only virtually. As fantastic as this sounds, such simulations may become reality, not created by Aliens, but directly ourselves. Artificial Intelligence gets more sophisticated, and if aims to imitate human-like behavior, even experts have problems in distinguishing it from human beings. A Metaverse platform allows users from all over the world to connect. Due to this, most users do not know each other from outside the platform. The more the Metaverse extends, or even include parallel worlds, the less the possibility for a user to meet other users. The relative loneliness may not be perceived by the user, as there are numerous other characters to interact with, in most of the cases controlled by an AI, which could be directly by the platform or other organizations, like for example external companies or other organizations. We would live in a simulation, not aware up to which grade it goes, as we may

---

[46] The Wachowski Brothers (1999, 2003): "The Matrix"

perceive other characters as humans. Of course, we still are aware of the simulation, as most the time we are still in our atom world, nevertheless, the boundaries blur and may overwhelm our senses.

10 THE AI COACH

The integration of Artificial Intelligence into companies will create completely new job profiles. One of them is the *"AI Compliance Officer"*, but also the "AI Coach".

After AIs already won in Chess and other board games against their human opponents, in 2015 Google let their DeepMind-AI learn different classic Atari games[47], such as *"Ms. Pac Man"*, *"Space Invaders"*, *"Video Pinball"*, *"Q-Bert"* and *"Montezuma's Revenge"*. The AI started with trial-and-error. The software learned based on *"classic conditioning"*, as it received a positive amplifier. This similar to *"Pavlov's Dog"*. In the beginning the AI made only slow progress, but then advanced and at the end could beat the human high scores in Space Invaders and Video Pinball. With the more complex Ms. Pac Man and Montezuma's Revenge the program still struggled.

Around two years later Microsoft changed the setup to *"cognitive learning"*. The AI observed human players and learnt from its mentors. In total, the human players created 45 hours of gameplay, which was analyzed by the machine. The AI still had its problems with Montezuma's Revenge, but in average the machine learnt faster from the human players, as it did on its own before. Based on this, Microsoft concluded that with adequate human teachers an AI is

---

[47] Cooper, Daniel (2015): "Google's newest AI can beat your Atari highs-scores"

learning faster.[48] Just as it applies for humans.

Inside an organization employees may require temporary coaches, as they have the required technical knowledge, but may lack of soft skills. Such raw talents need support by experienced colleagues to reach the next level of their career. A possible solution is to team them up with a higher manager, who acts as a coach, so that he / she can learn the required skills, such as emotional intelligence. Cognitive learning is used.

Like this scenario, special coaches may teach AI software to make the adequate decisions. This is relevant as decisions not only have to be maximized for the short-term, but to ensure sustainability to maximize the long-term profit. Ethics & Compliance must be obeyed, even if impunity would not punish violations to law. The AI must understand that nevertheless there is a cost of corruption, which can manifest itself, for example, in shrinking markets, raising costs and low profit margins. Even potential fines based on global investigations must be considered. Business decisions must be based on law and values. Furthermore, the AI must fit to internal the organization, including to the human employees. Based on region or even group culture, the software must interact differently with its human colleagues. Human and machine diversity are no single topics, but the human machine group requires such.

Human AI Coaches can teach such ethical decision making to ensure that the algorithm mathematically understands

---

[48] Dent, Steve (2017): "Humans can help AI learn games more quickly"

that transparent business ensures long-term success, even if on the short-term this may lead to lower results. As each company has its individual Code of Conduct, mostly based on its founder, decision making has its individual variations, so that the machines could not learn such behavior automatically from similar software used in different companies. Trail-and-error can easily lead to high fines and reputational damage, cognitive learning and "human understanding" is the more effective solution.

Today's Compliance Officers claim that ethical behavior leads to sustainability and is a sales advantage. The vision must be expressed in mathematical formulas and show a higher expected value than the output of corrupt behavior. If this is not possible to explain, not the AI, nor human employees will understand the message.

Microsoft's Maluuba AI was a good student and reached in Ms. Pac Man the perfect score of 999,999 points, more than every human player reached in history.

11 CATHARSIS

The Greek word "Catharsis" can be translated with "Cleaning". At the end of the 19th Century the noun was introduced into the world of psychology. With "Catharsis" the scientists defined the potential effect that with living out aggression, for example in boxing or other sports, the individual can reduce its negative emotions. First studies confirmed the theory, but later in the 60' to 80's most studies came to the opposite result that Catharsis fosters aggressive behavior.

In the middle of the 80's the theory made it a last time into the focus of discussion, as experts argued about the effects of warlike computer games, especially as they mostly got played by children and youth. Game publishers argued with the catharsis-theory and politicians often held the learning-theories against that, as the users of such games learned aggression as an accepted behavior leading to a positive result. At the end, both theories could not convince, as video games still not had realistic graphics and the low-pixel animations could not get perceived as human beings by the players. Further studies concluded that warlike games could foster aggressive behavior, if other negative circumstances are given, as non-functional family structures or peer pressures. Some countries as Germany took consequences and set several titles on an index, so that only adults had been able to buy them. Furthermore, advertising for these games had been forbidden.

Today experts confirm that video games can have a positive

effect on children, as reflexes get fostered. Gamification bets on the power of transporting information while playing. With the rise of the Metaverse, we have to come back to the learning effects.

The US-author Daniel F. Galouye published in 1964 the novel *"Simulacron-3"*[49], the first book describing a Metaverse-like Virtual Reality. A company developed a super-computer, which is able to simulate a whole city, until the smallest detail. Via VR, humans could temporarily stay in this second reality. Similar as in *"Total Recall"*[50], users of such virtual worlds have the possibility to take on different roles, including having super-powers. Also, it is a safe environment to live out aggressive behavior. In opposite to '80s video games, today's VR comes close to reality. The more real and virtual worlds blur and mix, the higher the chance that learnt behavior in the one world could get used in the other one. What is valid for the good, is a risk for the bad. If players stay a longer time in a perceived as real 1930 Chicago simulation with intact Mafia structures and corrupt politicians, corruption may get learnt as adequate and successful behavior. If individuals are not able anymore to distinguish between virtuality and reality, learnt behavior will be applied in both. A special risk for the Metaverse, where artificial and real individuals act together in the same space. A virtual Pokémon can be treated differently than a real-life dog.

---

[49] Galouye, Daniel F. (1964): "Simulacron-3"
[50] Dick, Philip K. (1966): "Total Recall"

12 MACHIAVELLI FOR AI

Besides its Latin name *"divide et impera"*, the expression *"divide and rule"* is not based on the Old Romans, but on the Italian philosopher, politician, poet, diplomat and historian Niccolò Machiavelli and his famous book *"Il Principe"* (Italian for "The Prince"), a book about the philosophy of politics and government, especially written for the Medici family, which ruled Firenze and had been the most important family dynasty at this time.[51]

Divide and rule stands for the principle to divide a big group into smaller sub-groups, to make them easier to rain. This not just through pure dividing, but also a construction that the new sub-groups are not having relations with each other, but the relations are limited to the one leader. A strategy, what was not Machiavelli's invitation, in fact he received the inspiration for this by the foreign politics of the Roman Empire, already used by Emperor Cesar himself.

Nearly 500 years later this principle gets used for Artificial Intelligence and its learning process. Developers at the Microsoft company Maluu used this concept to teach their AI playing Ms. Pac Man. The software divides the different tasks into sub-tasks:

- Eat energy pills

---

[51] Machiavelli, Niccolò (ca. 1543): "The Prince"

- Eat the additional fruits
- Escape the ghosts
- Catch the blue ghosts

As the AI (similar to traditional software) is able to multitask, it calculates for each sub-tasks its own learning algorithm, including virtual positive multiplier for successful executed sub-tasks (as eating an energy pill).[52]

The four sub-tasks work simultaneously, but completely independent from each other. The AI takes the average value as the direction for Ms. Pac Man. Just as Machiavelli defined, the AI separated the four sub-tasks and then evaluated on the next higher level the results to merge them into the decision in which direction Ms. Pac Man should move.

In times where we became Cyborgs, as part of our memory was outsourced to Wikipedia and other parts of the internet, we can analyze how machine learning may get used for humans, too. Companies are in the process to go find new paths for their training, and "micro-learnings" is one of them. Instead to do a complete training and explain all aspects of a process, the different tasks get separated and included in small five minutes online-trainings. As the trainings are short, employees stay motivated and concentrated. A series of micro-learnings replaces the

---

[52] Simonite, Tom (2017): "Microsoft Masters Ms. Pac-Man with a Horde of AI Agents"

traditional longer web-trainings.[53] If required, single episodes can get fostered and repeated individually. At the end human and machine learning is quite similar.

Translating these understandings to the Metaverse, the AI character of the Librarian functions as a personal teacher and counselor. Of course, this figure can give lectures in university halls, but its strength is to be directly with the user's avatar to counsel and give vivid short explanations. Furthermore, as the Metaverse maybe be different for each user, the Librarian does not have to attend several users but could attend them individually. This as each user may have a different Librarian, or even that each user will be active in a different manifestation of Virtual Reality.

---

[53] Henz, Patrick (2017): "Access Granted – Tomorrow's Business Ethics"

13 THE HEART OF AI

Artificial Intelligence is a software and due to this, acts on algorithm and pre-defined orders. Software and robots which can re-program itself are still science fiction, even though scientists are already experimenting with such solutions. Due to this, the character of the AI depends heavily on the programmer team, the "creators".

Based on this relation, this group is sensitive for the company. It not only requires the highest talents, but furthermore an effective team play. The team must represent the complete company, including all its values and diversity.

Different languages are a key factor for diversity. This as thinking influences language and language influences thinking. The philosophical question what was first, thought or word, is still not solved. What studies confirm is that multinational teams are more efficient than single national groups. Different languages not only consist of different vocabulary, but furthermore different grammar and structures. People coming from different culture perceive their environment differently. If such employees can work efficiently together, if confronted with a problem, the group will come up with more ideas how to solve this.

The benefit of language goes further, "speaking an additional language provides greater cognitive and emotional understanding than just the native tongue."[54]

With diversity and inclusion, the programmer group has an easier access to understand the requirements of the company and elaborate empathy not only for their needs, but also for other stakeholders, who will be directly or non-directly affected by the AI and related decision-making. High levels of diversity ensures that the code of the AI will not include cultural or other biases.

---

[54] Hogan-Brun, Gabrielle (2017): "People who speak multiple languages make the best employees for one big reason"

14 DATA PRIVACY FOR THE DIGITAL TWIN

Digitalization plays an important part inside Industry 4.0. Production and test facilities, but also office infrastructure can be simulated inside the computer. These virtual structures can run in parallel to the real-world locations and as the "Digital Twin" receives continuous information from its counterpart, the simulation gets more adequate.

The Digital Twin is used to analyze the efficiency of the complex system, especially to predict how it reacts to changes. An important tool to understand when and which updates are required, or for example, when temporary downtimes must be planned to install required spare parts.

The idea is not completely new, as Sid Meier's 1990 "Railroad Tycoon" is only one example for successful economic simulations, played on MS-DOS and the then actual Amiga- and Atari-home computers.

If we follow Edward Deming's philosophy, employees are part of the system. So not only the change of hard- and software will determine the output, but also the change of employees between the different positions. This based on skills and experience. If we think of other games, as sports management simulations, companies may not only include long-term skills into the digital twin, but also the actual performance of the employee, including medical leaves and the last results from the regular evaluations. Like soccer players or other athletes, corporate employees not constantly work on their highest levels, but are influenced by mood, distraction, health, and other factors.

The computer simulation is not limited to picture the machine or structures, but also could save digital twins of its employees. These characters can include all available HR information, so that an AI can analyze on which position the employee is most effective and furthermore predict performance.

But not only the HR department is a potential source of information, it can come also from automated devices. Today's voice-based interfaces are integrated into smart phones and operational systems. They achieved a higher independence thanks to intelligent loudspeakers, as Amazon's Alexa. Here the AI is not an additional app, but the device was designed around this software. The AI is connected to the Cloud, so that all requests get analyzed and even stored as big data. An industrial solution would send such data to the company server, where it may get connected to the HR-files or included into the Digital Twin. These interfaces could be connected to most machinery and thanks to this, Industry 4.0 solutions adapt to the user and not the other way around. Nevertheless, adequate skills are required. As the interface is not only connecting human with the machine, but furthermore is online with the server, it is capable to analyze the user commands, including its results. The human is completely transparent, and the company is always aware on what level of efficiency he, she or they perform(s).

To deliver the most accurate results, the simulation must

include the employee's long-term skills, but also the changing short-term conditions. As companies may give access to their external partners to use the digital twin, it is understandable that data privacy regulations are imperative. This to protect the individual's fundamental right of privacy and in addition to this, to protect the company against the possibility that an external partner would woos away talent. Digital Twins may be driven by the company's IT- or project management department. Normally, they are not specialized on data privacy or human labor-law. It is required that HR and Compliance accept the changing environment with its new roles and responsibilities. To limit legal risks, these departments must be included in the design, setup and execution of the Digital Twin, especially as different countries, or regions, as the European Union implemented more restrictive data privacy laws. Such regulations do not prohibit the usage of personal data but require from the organization to be transparent with the data it collects, and which algorithm predicts for what purpose.

Digital Twins are not limited on technical processes, but also can get used to picture white collar processes. Products like Microsoft's Office 365[55] can understand, based on emails and calendar data, communication patterns and give insights who in the office work frequently together. Based on such information, offices can get re-designed to facilitate communication, as grouping employees together and so bring communication from on- to off-line. In most companies, employees are allowed to use office IT

---

[55] Microsoft (fetched 08.11.2017): Microsoft Workplace Analytics

infrastructure also privately, so depending on local regulations, such additional insights may cause data privacy concerns[56], especially at it may also include externals.

---

[56] Tushman, Michael L. / Kahn, Anna / Porray, Mary Elizabeth / Binns, Andy (2017): "Email and Calendar Data Are Helping Firms Understand How Employees Work"

15 THE STRANGEST THINGS

Netflix's hit series "The Stranger Things" sends us back into the eighties. A time where videogames still had been at maximum two-dimensional. Space Invader's laser cannon could go to the left and right, it moved on just one dimension X. Pac Man ran through a labyrinth, he used already two dimensions, X and Y. In opposite to these electronic games, roleplay adventures a la Dungeons & Dragons became en vogue. Even if existed miniature figures, the game visualized inside the player's imagination, as a game master read from a book and explained the challenges, a real 3D-experience. In addition to the group games, there had been also published several solo adventures as books. The reader had the possibility to decide in different situations how the character should react. Depending on this, the adventure continued on a different page. Based on the combination of decisions, it was always a different story to experience.

Time is generally defined as the fourth dimension. If we define that it is possible to move free inside the dimensions, it would be possible to move in time from beginning to end, and end to beginning. Similar to rewind a VHS-tape. A general possibility but based on the limitation of our existence not possible for a human. Just as Pac Man is not able to escape his two-dimensional labyrinth. The consequence of a four-dimensional universe would be that the free will is an illusion, as everything already is decided. Due to this, humans would not be accountable for their actions and decisions.

If we analyze our life, sometimes small decisions had a big impact. If we had acted differently back in time, our today's life would be completely different. Actual theories indicate that there may exist a near infinity number of dimensions, so practically each decision would open up a new universe. The Big Bang was not only the start of a three-dimensional spreading over the time, but furthermore the unfolding of universes. Concluding this theory, there exists for each action a consequence and each possible decision had been taken in a different dimension. Ergo, the individual is free to choose and can be hold responsible. This philosophy underlines the rule of law, as the human is responsible for its actions, no excuse. As everything what can be imagined, happened in some far-away universe, we have to take care that Pac Man does not stand behind us.

16 Artists Needed

Today's technology offers numerous possibilities to disrupt today's business or life in general. Still lacking is often creativity to identify a real benefit. Virtual Reality glasses are on the market. Thanks to the Google Cardboard[57], such glasses can be acquired for very accessible prices, what makes them interesting also for schools and students. Especially for learning purpose, 3D videos are interesting, even without direct interaction. The learning effect raises with involvement. The virtual reality mediathek by the German public television broadcaster ZDF presents such an example, as one of their videos teleports the user inside the historic Roman Colosseum.[58] Important that there is no single position, but thanks to the glasses, the individual stands sometimes between the fighting gladiators and later besides the imperator. To follow the action, it is required to change the viewing angle to look in different directions. Due to this, the user cannot sit in chair, but must stand up, just as inside the virtual experience. If this is not the case, it is impossible to follow the plot.

The story gets not automatically presented but must be discovered. Due to human psychology, what is more difficult to receive, perceives a higher valorization. Information, which had been concluded by the individual itself, gets easier remembered than such, which had been directly presented. So, if used with creativity, VR glasses

---

[57] Google Cardboard (fetched 28.08.2017)
[58] ZDF (2017): "Gladiatoren im Kollosseum"

does not only can offer a more plastic experience, but also support the learning effect. Only one more problem, many individuals feel nausea while using these devices. This problem arises because of the disconnect between eyes and the inner ears, where the human sense of equilibrium is located. While using the glasses, eyes and ears receive different information, what the human body interprets this as nausea. So far, a relevant problem, which prevents the further commercial success of the VR glasses. So, no surprise that different companies work on the problem.[59]

Even if no learning effect is required, the artist can bring the technological possibilities to a higher level. This is comparable to painters as Pablo Picasso, who used the screen not to picture reality, but to create art. Based on this philosophy, Commodore Inc. invited pop art artist Andy Warhol to present in 1985 their new Amiga computer. In opposite to other machines, this 16-bit computer had the ability to show up to 4096 colors on the screen. As the Amiga was completely new, the engineers had not been completely sure if the software would run stable on the new machine. Luckily the presentation went well, and Warhol edited live a photo of the also invited Blondie singer Debbie Harrie. An impressive start for the Amiga, which should sell later more than six million units.

Today, creativity can be often found in the independent gaming scene. The big commercial studios invest millions of dollars to create new titles. Similar to the movie industry, this often blocks creativity, as the financial success is mandatory to ensure sustainability of the business.

---

[59] Metz, Rachel (2017): "Now There's a Nausea Dial for Virtual Reality"

Questions if to realize a new concept or to publish the successor of an already established successful game mostly gets decided for the last. Just like classic arcade titles, independent programmers can create art, without 3D graphics or professional soundtracks, but mainly based on a unique idea.

Smart phones apps offer additional possibilities, as the specific location can get included, even the direction, in which the user looks. New possibilities for the artists to infuse interactivity to the videos, as used by the Foo Fighters. Their app combined a performance of "Sky is a Neighborhood" with a view on the user's starry sky.[60] Underlying the song's message, the app gave the user additional information to the constellations. This knowledge supported the feeling of a cosmic neighborhood.

With this, put on your VR glasses to experience, "A Day in the Future" …

---

[60] sky.foofighters.com (2017): "Sky is a Neighborhood"

17 GAMIFICATION^2

Gamification is mostly seen as a possibility to communicate information. But the development went already one step further. The prestigious University of Geneva will partner up with the science-fiction online game EVE. The university will provide 167,000 deep space light curve images to the game community, where the users can use their virtual spaceships to explore this information and support science to discover new exoplanets. EVE Online includes up 500,000 players and due to this, presents a relevant source to support the chronically understaffed scientific projects.

The game simulates a virtual world, where players can take on the role of spaceship captain and discover the wonders of the galaxy. With the connection to the university, the game becomes reality, as the users, sitting before their home computers, become scientists acting inside a virtual space. The software simulates a virtual object based on the scientific data and, inside his or her role, the human user evaluates it. If enough users evaluated the planet information, the games send the information back to the university.

Back in 2003, EVE Online was designed as Massively Multiplayer Online Role-Playing Game (MMORPG). Today even conventional games offer an additional multiplayer option, where the users can play against others, independent where they are in the world. The Cloud makes it possible. It is up to the game-designers to explore new possibilities for this technology. But the trend that games and reality grow closer together will continue, and this not only because of the advancing graphics.

Thanks to the continuous connection to the internet, real-time information can be included into the gameplay. Users could play Soccer against professional teams from the actual game day, including details as temporary excluded injured players or the actual condition of the team. Furthermore, the simulation can take the local weather conditions into the account. If it rains outside the window, so it does inside the virtual stadium. Such additional conditions bring the game nearer to reality, as the player does not react in an independent space but stays connected to the outside world.

The US manufacturer of wrist-worn straps that measure health information Whoop has an agreement with the NFL Players Association, which enables participating players to sell their health data to the company. Worn fitness devices connected to the Cloud are an ideal possibility to include real-time information into videogames and other apps. Ethically questionable, as based on the contract, players are free to participate or not, but nevertheless the access to biometric data may be a requirement to join a top team, so that the individual player's possibility to freely decide about his or her personal information is limited.

In opposite to "Pokémon Go" what presents "Augmented Reality" with including virtual characters into the real world, such technology would go the other way around, with including real circumstances into the virtual world, due to this it could be called "Augmented Virtuality".

Besides these possibilities, the trend may also get a dark side. Similar to the classic migrant workers of the past, "Crowd-workers" may be a similar group in the near future. Artificial Intelligence will replace many white-collar jobs, especially jobs with a high part of repeating tasks. These skills devalue. The replaced employees have the choice to develop new skills or offer their existing ones for a lower salary. One example for the last is that organizations will hire personnel providers for single tasks. The subcontracted employees work from home, using their private computers, connected via a virtual network to their temporary employer. Similar to today's UBER or even traditional Pizza taxis, the crowd-workers offer not only their-selves, but additionally their work-tools. What for the UBER-driver means the private car, is for the crowd-worker the private computer. This way, AI does not replace the human employee, but devalues them so far so that their usage will be cheaper than replacing them with the software. This at least until the next level of technological achievements. A potential scenario, as Stephen Hawking warned that AI and Industry 4.0 is going to reduce typical middle-class jobs.

UBER is already testing driverless cars and pizza delivery drones are in experimental phase. To avoid the described development of human value, different politics, but also business experts as Bill Gates and Elon Musk started a discussion to discuss a universal income, where individuals receive income from the government, independent if they work or not.

A first type of crowd-workers already exists. So far based on personal decision and part of a lifestyle. Digital nomads are individuals, which use modern telecommunication as internet and cloud to work from remote locations. With these possibilities they finance they personal interest to travel the world. As positive side effect, such temporary locations require lower costs of living than in their home countries, as mostly Europe and the US. Nomadism can further used to reduce taxes, as without having a fixed home-address, individuals may be able to construct legal tax-saving-models as open up micro companies and use offshore bank-accounts. As these modern nomads understand their business model more as lifestyle, the locations are an emotional choice. Based on cost, internet-connection, fun and safety, Berlin (Germany), Bangkok

(Thailand), Budapest (Hungary), Ho Chi Minh City (Vietnam) and Chiang Mail (Thailand) are trending. Digital nomads are, in most, a sub-group of Gen Y and Z. Due to this, potential negative consequences, such an insufficient healthcare and insurances, are not measurable yet.

Unnoticed digital nomads became part of company structures. Various organizations implemented open offices with paperless desks. As result, employees produce no papers, as there is no space to store them. Often, they not even have a fixed place and sit each day on a different desk. Even if the personal contact with colleagues is still needed, working days from home or other locations is possible. So even if employees continue on the payroll (and are protected by local labor laws), many of us became already a digital nomad.

18 WHIZ KIDS

The implementation of new media always got accompanied by the fear that it would produce a negative impact, especially on the younger generation. Similar happened with TV, the first video-consoles, home computers, internet, and smart phones. The actual Generation Z are digital natives. Since their first steps, they are accustomed to have touchscreens and information around. As parents used tablets and smart phones to calm down their toddlers, they learnt early to use such devices and even got confused if they cannot change the channels on TV with the same whishing movements. In opposite to Y, Gen Z in most cases even had been socialized with AI, the various voice assistants. Due to this, it came by no surprise that the report "A Future with AI: Voices of Global Youth Report" concluded that young people (10 – 24 years, coming from 36 countries) stated in 80% of the cases that they interact with AI multiply times a day. Doing so, they were not uncritical, as they were aware of the risks the technology poses, but nevertheless, mostly (93.2%) perceived AI and robots as positive, and in a lower percentage (68%) as trustworthy.[61]

All actual studies confirm that Z has strong values and expects such also from their potential employers. In opposite to the 80s' YUPPIES (Young Urban Professionals), salary lost parts of its importance. This as

---

[61] Hobenhout, Lambert and Takahashi, Toshie (2022): "A Future with AI: Voices of the Future"

status symbols changed. Expensive cars and apartments (also as such may be perceived as out of reach) declined in importance, but in opposite, fast internet connections and virtual acknowledgment, via social media, blogs, and 3D world building games, gained relevance. In his study "Growing Up Digital", Tap Tapscott concluded that these generations are *"smarter, quicker and more tolerant of diversity than their predecessors."*[62]

But there is one downside about the younger generations, their respect to data privacy and content copyrights is significantly lower than at anterior generations. Their different socialization explains it.

Secondhand Shop, Aachen, Germany

Products are more than their tangible part as they also

---

[62] The Economist (2017): "The kids are alright"

include an emotional universe. This is today not only relevant for industrial designers, but especially for the developers of digital content. More and more music, books, videos, software, and all kind of art (as offered via NFTs) are not bought anymore in a physical store, but directly downloaded to computer, mobile phone, tablet, or TV. As in the past, a detailed user manual or artistic cover had been part of the complete package, the consumer had something in their hands to conclude from this to the quality & value of the intangible content as music, video, or software. As this is missing today in many cases, people lost the respect for the product and piracy is often perceived as a face-less crime, as nothing gets physically stolen, just additional copies elaborated.[63]

Gen Y and Z show a similar mentality also for their own information, as they often present their whole life on Instagram, blogs and/or Tik Tok. Non-Fungible Tokens are designed to counteract this tendency, as they make virtual objects (which could be or not be connected to a physical object) unique and potentially non-copyable. First adaptors starting spending money on virtual goods, which could be a unique piece of art, but also virtual objects for games, or clothes for an avatar. Children born since 2011 form the latest Generation, called Generation Alpha. A lot of its members are in the process to grow up and getting socialized including Metaverse-like platforms, such as Fortnite or Roblox. They spend money on such games, but so far, only the one earned by their parents. It has to be observed, if in later years they will show a higher tendency to not only spend time in the Metaverse, but also income.

---

[63] Henz, Patrick (2016): Business Philosophy according to Enzo Ferrari"

19 MORE TROUBLE WITH BUBBLES

Amazon's Jeff Bezos bought the Washington Post in 2013 with the promise to not intervene with the newspaper's editors but leave the paper independent. So far, he kept his promise, and the Post is on the way to keep up with the times. In 2016 debuted "Heliograf", a new software which supports the automation of articles. The software can connect to data bases and take results, as for example from elections and sport events. This works not autonomously, but the journalist must feed Heliograf with narrative templates and key phrases. Due to this, the automated articles still feature the journalist's individual style. Doing so, today's mission for Heliograf is not to replace the human employees but ensure an effective usage for them. Today's world creates every day numerous small events, important for a small number of involved people. They could be locals or also globally distributed. As the public interest is limited, it would not be possible for newspapers to cover each of such events by a human employee. Accordingly, to this vision, Heliograf's first job was to cover the first-round events at Rio's 2016 Olympics, what the software did without problems. Later in November 2016, it wrote 500 articles with such limited human intervention. That way it created 500,000 clicks on the Post's portal, what is again relevant for the newspaper's revenues.[64] The paper was completely open about its artificial author and, for example,

---

[64] Keohane, Joe (2017): "What new-writing bots mean for the future of journalism"

publicized this pro-actively via their Twitter-account "@WPOlympicsbot".[65]

The newspaper underlined that the software would not change the Washington Post's political attitude or its philosophy. Nevertheless, its competitors may push the development further. For the information flow are sender and recipient relevant. Heliograf automated the sender. A next-generation software may not only connect to information databases but also analyze the online behavior of the user. Cookies and apps collect user information, due to this, an intelligent software can create a profile and forecast not only what topics are interesting for the individual, but also what it wants to read. This based on the idea that the user not only wants to get informed, but also to confirm its already existing opinions. This makes it possible for the news-portal to create tailor-made articles for every user and everyone would see a different page. Covering different events and even if they would cover the same event, the text would vary.

The individual stays in its personal comfort zone and potentially only sees the information and opinion what it wants to read. But this avoids the reception of alternative ideas, which could challenge the person to re-think the existing opinion. If the user does not actively seek information on other news-portals, the person stays in the existing information bubble. On a higher scale this means that the division of society continuous and the different

---

[65] WashPostPR (2016): "The Washington Post experiments with automated storytelling to help power 2016 Rio Olympics coverage"

groups inside the region even may drift farer away.

For populistic governments, this means good news, as a political leader can use *"divide et impera"*.

Stephen Hawking once should have said: *"The greatest enemy of knowledge is not ignorance; it is the illusion of knowledge."* Due to bubbles, the users receive biased information. Often without being aware of, only "one side of the medal" gets perceived. As the individual does not know that information is missing, it falsely assumes that everything is transparent and based on this, make its decision.

The Metaverse has the possibility to multiply bubbles. As discussed, algorithms can create AI controlled avatars tailor-made for the user, including outer appearance, behavior, voice, attitudes, etc. An avatar which may appear similar to the user, would most likely be perceived as sympathetic. If first interactions which such an AI character would be beneficial for the user, sympathy evolves to trust. If such a level gets reached, the AI is able to influence up to manipulate the human user. The more information the algorithm has about the user, the easier to perform such mimicry. It does not have to stop with the avatar, if enough computer power is given, the algorithm can create for each user a different appearance of the Metaverse. Depending on the user and the platform's approach, this could be a place for adventures, but also an environment to keep the individual in its personal bubble.

20 A Different Show

The 1998 movie "The Truman Show"[66] presents us a tv-show where everything is around Truman Burbank, a potential average person, who is unaware that he lives in a fictive town, with actors all around him, and millions of spectators following his daily life. The Personal Digital Twin includes information about the user, an AI for decision making based on perceived stimuli, and the avatar. All what is needed to take the PDT and include it into different scenarios. The result could be used for scientific experiments, training, or pure entertainment. Due to limitations of the algorithm, the PDT is uncapable to analyze that it is inside an observed scenario and acts purely to the perceived stimuli of the situation.

This works also into the other direction. One day, human users can use their avatar to enter pure entertainment scenarios, as joining a Metaverse version of "The Sims", joining a family or other group of AI characters. If ethical regulations will not demand that AI must disclose their artificiality, as today expected from service chat-bots, computer generated Sims-characters may be undistinguishable from human players.

A look on popular multiplayer games may lead to the conclusion that it will not stop there, but most likely include fantasy roleplay, science fiction and shooters. Author and director Michael Crichton gave us his vision with the 1973 movie "Westworld".[67] Inspired in the different sections of

---

[66] Weir, Peter (1998): "The Truman Show"

[67] Crichton, Michael (1973): "Westworld"

the Disney Parks, the movie presented a futuristic theme park, where robots were not limited to mechatronic puppets, but had been autonomous machines, in appearance and acting indistinguishable from human visitors. The fictive 1983 US high-tech park featured three sections, the old Wild West, a romanticized Medieval Europe, and a decadent vision of the Roman Empire. Of course, robotics not developed like this until the 1980s, and still in the 2020s, robots did not reach such levels as shown in the movie. Experiences like promised in Westworld could be provided via the Metaverse, including those limitations that the robots had not been allowed to kill the visitors are not needed anymore, as only avatars are affected.

Crichton retook his theme park idea and published in 1990 his novel "Jurassic Park",[68] which only three years later had been released as movie.[69] No need to say that a Jurassic Park in the Metaverse would be much safer, as it would prevent the risk of spreading Jurassic life outside the artificial boundaries, or as the fictive Dr. Ian Malcom said: *"Life finds a way."*

---

[68] Crichton, Michael (1990): "Jurassic Park"

[69] Spielberg, Steven (1993): "Jurassic Park"

21 Risk & Compliance in the Metaverse

Big Tech companies bet on the Metaverse, a Virtual Reality combining the various services already found on today's internet. Typically, users can navigate through this VR via a 3D representation: the avatar. With this, it is more than just a social media platform or communication tool but aims to be one VR connecting all online services.

Even if still far away from the described visions, some few global companies already use internal social media platforms to support the internal communication, exchange of information, up to let employees connect as self-created working groups. Separated from internal social media or 3D meeting tools exist knowledge platforms, chat-bots, or the HR-database. Also, Digital Twins are already an established concept. So far, such get used to monitor existing machinery or plan updates or completely new hardware. It is up to the Metaverse to combine all these existing platforms to one holistic one. Here, employees can not only walk around the graphic figure of the Digital Twin, but thanks to avatars and VR glasses directly operate the machinery.

Nothing can replace in-person workshops, nevertheless training inside the cyberspace can add to this. Various Compliance risks have a high impact, but a low likelihood. To prepare employees for such risky scenarios, VR can be used to create such cases and let employees act inside this simulation. This can be perceived as a safe sandbox to not only try the learned adequate behavior, but also to try out undesired behavior to see what the potential consequences for him or herself are, the company and society. Such cases would be like 1980s role-play games as widely described in "Ready Player One". A fascinating opportunity for companies to enrich its company training, but also including pitfalls. It must be ensured that employees can create an avatar, from which they felt adequately depicted. Furthermore, AI fueled virtual characters must be as realistic as needed and avoid showing cliches.

Depending on the employee's job position, individuals can perceive different scenarios, like corruption at the immigration, customs control, or project site. Algorithms cannot only simulate potential scenarios, but also combine observed behavior with employee database and risk location maps. The result could be calculations of potential vulnerability of employees deviating to regulation and guidelines; similar as seen in "Minority Report." Of course, this is questionable considering ethics and data privacy regulations.

If not used for predicting an individual's career, the algorithm may create a video of the predicted behavior versus the required behavior. This way, the employees can perceive themselves from the outside, see them practically as a separate person. Assistant professor of psychology Rachel White explained: *"Self-distancing gives us a little bit of extra space to think rationally about the situation."*[70] As perception as a separate person reduces emotional engagement, employees are able to logically analyze their behavior. Furthermore, a comparison with a situation where the same character acted correctly should make them understand that they have the capability to do so, independent if it is in the Metaverse or the physical world. This especially as adequate behavior, as for example denying a bribe, stop bullying, etc., should not require being a super-hero. Employees could see a traditional video or a Metaverse 3D recording, where they could ghost-like move through the scenario, but not alter it. Seeing themselves first failing, but then mastering the

---

[70] Robson, David (2020): "The 'Bat Man Effect': How having an alter ego empowers you"

scenario thanks to a change of choices and actions.

For most organizations, such holistic concepts are still future music. Nevertheless, two developments lead to the corporate Metaverse.

1) Generation Z is entering the workforce. This generation has been socialized with the omnipresent access to VR platforms, like for example Minecraft, Roblox or Fortnite. Due to this, they do not perceive VR as separated reality, but part of a mixed reality, where atomic and virtual reality form one interconnected reality. As conclusion, actions in the VR impact the physical world. Consequently, Gen Z spend money for virtual goods, up to non-fungible tokens (NFTs).

2) COVID 19 accelerated the already existing trend to Digital Transformation, which not only means to implement new technology, but also to use existing one more efficiently. In parallel, hybrid and full remote had been implemented to adapt to employee requirements and keep the company attractive for new and existing talents.

Accepting the concept of one holistic reality, it is clear that in the virtual world not only must exist the same regulations as in the physical world, but that the same consequences apply. As for humans blurry the frontiers between virtual and atom reality, rules must stay the same to not confuse the employees.

The US research company Gartner predicts that by 2026, 25% of the population will spend at least one hour per day in the Metaverse. This includes those parts of daily life, for example work, education, but also leisure will shift from the atom to the virtual reality. Accordingly, corruption and fraud risks will shift, too.

Unappropriated gifts can now include a NFT artwork, an expensive virtual car the customer's favorite racing game or the invitation to a restricted live event in the Metaverse. Money laundry inside online video games already had been observed, especially as many platforms support and encourage the players to buy virtual goods. Based on a 2019 survey by the Anti-Defamation League, 86% of online games players received threats, stalking, harassment, and hate in general. Companies must adapt to this change in perceived reality to ensure that the relevant functions (for example HR and Compliance) stay relevant trusted advisors.

INTERLUDE: ELEPHANTS ON STILTS

I hear strange footsteps behind me and look around. Two tiny elephants walk through the hall. Like miniatures, realistic, only with one difference, their legs are thin and much too long, about three times the size of the rest of the animals' body. Instead of the robust perception of a known elephant, these creatures look elegant, but furthermore fragile. Carefully they keep on walking towards a shelf. As if physics does not apply for them, they go right through the wall.

22 A SURREALISTIC METAVERSE

The name "Surrealism" had been coined by André Breton and Philippe Soupault: *"Surrealism is based on the belief in the superior reality of certain forms of previously neglected associations, in the omnipotence of dream, in the disinterested play of thought."* Surrealism is no absolute state, but temporary, as famous authors and artists switched between realism and surrealism,[71] like today's user between physical and virtual reality.[72] To fully use the given technological possibilities, designers may have to forget "Ready Player One", as even if it relates to early Dungeons & Dragons roleplay games, it still describes a Metaverse with valid physical laws, which do not apply inside a computer vision. Inspiration could come from Surrealism, not only the well-known paintings by Salvador Dali, but also the underlying philosophy, as defined by Breton and Soupault.

The Dali Museum already offered a first look on how such a surrealistic experience could look like. "Dreams of Dali" had been an immerse experience, where visitors of the museum could dive into the painter's world, using Virtual Reality glasses. Immersive yes, but not interactive yet, as the Metaverse would be.[73]

Using all these possibilities and visions to create the future Metaverse would distinguish this concept from the Digital City, which would be like a Digital Twin be a realistic, but enhanced, copy of the user's hometown. Based on Professor Derrick de Kerckhove, this with the aim to

---

[71] Breton, André (1924): "Manifeste du surréalisme"
[72] Henz, Patrick (2022): "The Societal Impact of the Metaverse"
[73] The Dali Museum (2019): "Dreams of Dali"

simplify services in the physical world, like for example administrative procedures, house management or the execution of direct democracy. The Digital City would cover basic and advanced requirements, while the Metaverse would satisfy social and self-realization needs.

Back in 1972, a Canadian college student asked the author Philip K. Dick for his definition of reality. After taking some time to think about an adequate answer, he replied: *"Reality is that which, when you stop believing in it, doesn't go away".*[74] A clear answer, but as it is limited by the dimensions, an open playground for philosophy.[75] On a more practical level, reality includes a certain kind of stability, as even if you leave your home in the morning, it most of the cases, it is still there in a very similar condition when you return in the evening.

AI art generators like DALL-E, DreamStudio or Midjourney create unique images based on keywords entered by the user. As these Artificial Intelligences continuously include additional data (images as found in databases and the internet), even if a user enters the exact same keywords in later point of time, the image will be different. The pictures can imitate different art styles or being photo realistic. Nevertheless, their creation only needs seconds. Such technology can be used to create for each user a unique perception of the Metaverse. Technology enables Surrealism, as the Metaverse would be unstable. Each time a user would visit a location, it would be completely different, including those locations change. For example, if users would frequent less a particular online

---

[74] Popova, Maria (fetched 1.3.2018): "How to Build a Universe: Philip K. Dick on Reality, Media Manipulation, and Human Heroism"

[75] Henz, Patrick (2021): "Tomorrow's Business Ethics: Philip K. Dicks vs. W. Edwards Deming"

shop, its location may automatically move more far-away from the user's home.

Due to its instability, the Metaverse will never be able to replace the physical world, as humans perceive the inner need for stability. Most people will perceive their comfort zone inside the physical environment, while the Metaverse will be a space for exploration and adventure.

22 Mount Olympus

Mount Olympus is not only the highest mountain in Greece, but also the legendary place where the gods lived, with a height of nearly 3,000 meters unreachable for humans.

The former Google employee Anthony Levandowski, for example in the development of autonomous vehicles, founded in 2017 the "Way of the Future", a self-proclaimed AI church, which had the aim to prepare mankind for the coming rule of the algorithms. Later in 2021, the world's first AI church closed its virtual doors.

The quote *"Mathematics is the language in which God has written the universe."* is commonly attributed to Galileo Galilei, even if clear evidence is lacking. Nevertheless, if would follow this idea, where else we would be looking for Mount Olympus today than in the Metaverse, a place where the laws of physics are not valid and only is limited by our imagination.

How different machine imagination from human visions could become shows the public accessible Midjourney algorithm, where users can enter words or sentences to describe an image, and the AI draws in seconds an image. [76] First examples of the Metaverse will still be created by human artists, in future, appearance may be directly created by the AI. Before this could be offered on a brighter commercial base, copyright questions must be decided. If a user enters a number of 10 words and the AI creates a picture or Metaverse, copyright holder is the human who entered the words or the company who offers the Metaverse platform? What if a user uses several pages from a Harry

---

[76] www.midjourney.com (fetched 02.09.2022)

Potter book to define a Metaverse? Returning to the beginning of the chapter, theoretically a user could enter the New Testament to define a Metaverse, including based on description program the various characters. When technical possible, some group will create this, as with "The Holy Land Experience" existed also a biblical theme park, ironically not too far away from Disney World.

What about the Metaverse being more hell than promised heaven? In Zoroastrianism, hell is described as a dark and lonely place, where the sinner must confront the Evil Spirit and get mocked by demons.[77] Even if the AI itself may not be evil, companies deploying it, may focus on the organizational benefit, instead of the benefit of the individual. Behavior inside the Metaverse is easier to monitor than in the physical world. An advantage for personalized marketing, which can continuously "mock" users, including manipulate them.

---

[77] Encyclopaedia Iranica (fetched 15.09.2022): "HELL i. IN ZOROASTRIANISM"

23 The Tron Metaverse

So far, when discussing the soon-to-come Metaverse, nobody is talking about "Tron",[78] the 1982 movie, which received in 2005 with "Tron: Legacy"[79] a worthy sequel. But are we really so far off this Disney fiction? As concluded in earlier chapters, the Metaverse may be populated by AI avatars. Today's voice assistants like Apple's Siri or Microsoft's Cortana (ironically named after a character from the computer game "Halo"[80]) are nothing else than a user-interface of a search-engine. Adding avatars to these algorithms would be a logical next step to include them into the Metaverse. That way users could directly ask these avatars to receive the required information. Inside the world of Halo, Cortana is an AI which manifests as hologram. To support the image transfer from the game to the Microsoft Phone, in both cases voice actress Jen Taylor gave Cortana her voice.[81] Also from Halo, Cortana already has a potential outer appearance, a design which could be used for an AI avatar inside the Metaverse. Adding a body to the algorithm may enhance the level of trust on the user's side. Similar to a discussion between two humans, the AI would not only passively answer to questions, but aligned with the concept of Cognitive Digital Twins, pro-actively inform about other relevant information, remind to appointments and even schedule new appointments for the user. Of course, such a development includes ethical risks, as more human the algorithm appears, the easier for it not only to influence, but also to manipulate.

---

[78] Lisberger, Steven (1982): "Tron"
[79] Kosinski, Joseph (2010): "Tron: Legacy"
[80] Bungie (2001): "Halo"
[81] Alhadeff, Emily (2015): "CORTANA – The smartest AI in the universe is more human than you think"

Continuing with this idea, what about AI driven avatars for other apps? For example, we can imagine a Microsoft Office assistant, Spotify DJ, Peloton Instructor, or Instagram Influencer. We can dream of a Light-cycle race between a Virus- and Anti-Virus-program. Why not?

Inside Virtual Reality, AI characters could be undistinguishable from the avatars by other human users, especially if the last may act autonomously in the case that their user is away from the computer. As Governments understand the risk that advanced technology may fool or even manipulate humans, various laws require that bots disclose their artificiality. An example for this is the California Bot Disclosure Law. Defined in paragraph 17941:
"It shall be unlawful for any person to use a bot to communicate or interact with another person in California online, with the intent to mislead the other person about its artificial identity for the purpose of knowingly deceiving the person about the content of the communication in order to incentivize a purchase or sale of goods or services in a commercial transaction or to influence a vote in an election."[82]

An AI character is an evolution of a chatbot, a three-dimensional user interface. This would apply also for AI characters, which do not directly assist a sale, but are designed to keep users on a commercial platform. Also, it applies for AI characters, which function as interface with a commercial application.

---

[82] California Legislative Information (2018): "Bot Disclosure Law"

Into the Metaverse

Cortana in Halo. Microsoft press photo.

24 The GONK Risk

In earlier chapters we discussed the possibility to act in the Metaverse with a sidekick, acting as an interphase to a search-engine or the platform itself. They could manifest, for example, as an omniscient librarian, a smart female hologram or a loyal dog.

What about the scenario that a hacker or organization "capture" our sidekick and would use it to subconsciously influence us?

US-psychologist Walter Bradford Cannon concluded 1915 in his theory of "fight or flight" that a high level of adrenaline can block the individual's ability for logical thinking, this as the hormone increases the heart rate to enlarge blood pressure, expand the air passages of the lungs, enlarge the pupil, redistribute blood to the muscles and alter the body's metabolism, as to maximize blood glucose levels.[83] Thanks to evolution, adrenaline acts as a motivator. Produced in low levels, it triggers to start cognitive processes. In high levels it provokes that the individual perceives to be in a survival situation and the primary task is to get out of this as soon as possible. Originally required to escape from angry mammoths.[84]

Based on this theory, low levels of adrenaline are required to start getting suspicious. If hackers could channel their message through the trusted sidekick, the user might be

---

[83] Becrzi, Ivan: "Walter Cannon's 'Fight or Flight Response' – Acute Stress Response"

[84] Henz, Patrick (2018): "From Auto Racing to Behavioral Science"

vulnerable to detect the potential influence.[85] This effect may get stronger if the outer appearance is rather unsuspicious. Users may never completely trust the omniscient librarian, as this character gets perceived as intellectual superior, same is valid for a character like Siri or Cortana. The dog is "the human's best friend". Intelligence is comparable to a 2-year-old toddler. Due to this, humans normally do not perceive a dog, especially if it is a puppy or smaller bread, as menace.

Between 100,000 and 15,000 BC, the first curious wolf decided to get into closer contact with humans, attracted by the prospect of food. The result became a classic symbiosis. The wolf gave up its freedom for regular food, while humans had to share their limited food but gained a guardian for themselves and their livestock. The creature advanced from the first to the third level of Maslow's 'Hierarchy of Needs'. Dogs acknowledge the human as 'alpha' and assume the role of 'beta'. They not only became an important guardian, but also a source of stress relief.[86] Although they act as the 'beta', thanks to their perceived cuteness, they are capable to influence their 'alpha', for example to receive attention and especially food.

In the Metaverse, appearance is just a question of graphics, something that human must unlearn from millions of years of evolution. Sidekicks, who appear to have lower IQ, aggression strength may easily get underestimated. This independent, if such AI character appears as dog with cute puppy-eyes or the famous GONK droid from Star Wars.

A world, where machines with different levels of intelligence co-existed already had been presented by the

---

[85] Booth, Serena Lynn (2016): "Piggybacking Robots: Overtrust in Human-Robot Security Dynamics."
[86] Henz, Patrick (2022): "The Human-AI Team – A Dog's Tale"

original 1977 Star Wars movie.[87] Of course, C-3P0 and R2-D2 had been part of the lead characters, but George Lucas created a big diversity of droids, and not all of them spoke fluently over 6 million languages as the golden robot.

Especially the power droids became a fan favorite. They are not a classic droid, more a mobile power generator. Due to this, its Artificial Intelligence was limited, what included its communication skills. The machine was only capable to produce "gonk"-like sounds, what became the droid's nickname.[88]

These GONK droids never played a main role, but often could be identified in the background. Another option to influence the users on a Metaverse platform, not use the sidekick, but let the message be communicated by some of the anonymous characters around.

---

[87] Lucas, George (1977): "Star Wars"
[88] Henz, Patrick (2017): "Internet of Things – The Gonk Risk"

25 Sitting down with Pablo Picasso

Finally, apart from the librarian someone else inside the library. A somehow familiar face, sitting on one of the tables. A glass of French red wine and a book in the hand. I look around that the walls, no more scrolls, just books. The next moment, no more books, I was inside a small and relative dark restaurant. Somehow, I know that the year is 1964, and I am here to meet the famous Pablo Picasso to speak about thinking machines. *"But they are useless. They can only give you answers."* He looked up his book as he was giving me his opinion.[89]

Questions, I have; answers, I received. What he wants to tell me? Do I have the right questions?

Assumed that the Metaverse will manifest like described in this book, or in a completely different way, what would be its impact? Maybe we spend less time in the physical world, and more in the Virtual Reality. Nevertheless, we may assume that in fact, we will be less in the grey zone in-between. Spending less screentime with classic social media and instead move inside the Metaverse, independent if physically we continue staring on a screen or use VR glasses.

The Metaverse, already in its early versions is, and will be, a form of escapism from the physical world. With this understanding, it is no revolution, but a continuous evolution, which started with the first humans. A voyage which began with storytelling, continued with cave-paintings, books, TV, movies, videogames, and social-media. The Metaverse combines these strings with the

---

[89] Fifield, William (1964): "Pablo Picasso – A Composite Interview"

physical part of Escapism, like paintings, sculpture, parks, or even entertainment parks.

The Metaverse has the ability to inspire its users, develop their human skills like creativity and ingenuity, on the other hand, it may also be designed to imprison people in their bubbles and comfort-zone.

Picasso took another sip from his glass of wine. Wine; wine is another form of escapism from the physical world. *"Maybe the Metaverse is useless, too."*

TOMORROW'S BUSINESS ETHICS: PHILIP K. DICK VS. W. EDWARDS DEMING

Many of Philip K. Dick's books and short stories seem like hallucinations, leading us to distant future scenarios, or into surreal parallel worlds. On a second look, we discover that these social and technical developments are emerging, if they do not already exist. Dick's philosophy is in harmony with W. Edwards Deming's System of Profound Knowledge. Both ideas jointly together show important ways to prevent a dystopia and to use the new technologies, such as Industry 4.0, Artificial Intelligence, Machine Learning, Digital Twins, Chatbots, Augmented Reality, Cobots, Robotics, Cloud, Autonomous Cars, Smart Data, Digital Transformation, etc., to humanize the corporate world, including a focus on ethics, compliance, and sustainability.

The COVID-19 pandemic acted as a disruptor and, in most cases, intensified the already existing currents. Action is required!

An entertaining trip into the present and future, ideal for Management, but also for other key functions such as Ethics, Human Resources, Strategy, IT, Governance, Risk, Compliance, Sustainability, including their holistic concepts such as GRC or ESG.

"Tomorrow's Business Ethics: Dick vs. Deming", 2021, 3.edition, 368 pages, ISBN-13: 979-8746471307. Also available as audiobook.

ABOUT THE AUTHOR

Patrick Henz started his career in Corporate Information and Compliance at the end of 2007, when he was responsible for the implementation of an Anti-Corruption program in Mexico and several Central American and Caribbean countries. Together with these tasks, he gained valuable insights into global Compliance programs, with a focus on Latin America. Since 2009 in his role as Compliance Officer he is responsible for an effective Compliance program; based on identification, protection, detection, response & recovery and combined with integrity, respect, passion & sustainability. With these means, he defines Compliance as pro-active function, being perceived as guardian, expert, and facilitator. The focus is on information to ensure adequate behavior, not only of the human employee, but Artificial Intelligence included.

He is member of the steering committee of the "IEEE Digital Reality Initiative", which focuses on various topics, including Digital Twins and the Metaverse. Furthermore, he is guest editor for the "IEEE Internet Computing Special" Issues on Digital Twins, deputy editor-in chief for Springer's Discover Artificial Intelligence Journal and co-chair at "MedicReS AI 2023 International Congress on Good Artificial Intelligence Practice & Innovation in Health Sciences 2023".

Since 2013, he lives and works near Atlanta, USA.

www.ingramcontent.com/pod-product-compliance
Lightning Source LLC
Chambersburg PA
CBHW031444210526
45464CB00005B/2321